AN INTRODUCTION TO NUMBERS

A KEEHN WORKBOOK

BOOK 1

by Michael Keehn

This workbook belongs to: _____

INTRODUCTION

This workbook is intended for the learning of mathematics, it is not a workbook targeting the passage of some home school or other bureaucratic test, however, such tests are much easier to pass when the student has a solid working knowledge. This book recognizes that students become ready to learn at different ages and that age has little to do with a student's readiness or interest in learning, thus the book does not target a particular age or age group. Therefore, this book may be appropriate for a child of six or a child of eleven or twelve years of age. Any student needing help in the basics can benefit from this workbook.

AN INTRODUCTION TO NUMBERS is exactly that. At the beginning, the student is introduced to the numbers zero (0) to nine (9), the foundation of all numbers in our number system. This is followed by a focus on zero (0), the center of the number system. The student is then introduced to the concept of a *number line*.

The student then finds that numbers are used almost everywhere to keep track of things. Numbers are used in baking, sewing, farming, electrical power, banking, retail sales, biology, almost everywhere we find numbers.

This is followed by counting, first by one, then by two, then by five, then by ten, then by twenty and finally by twenty-five. After counting the student learns the basics of addition, and following right on the heels of adding is subtraction. The existence of *negative numbers* is revealed, but this workbook does not provide substantial training in negative numbers, only an opportunity to know of their existence.

The workbook suggests group participation, working in groups of two or three, especially in counting. Each student in turn is to say the next number in the counting sequence, giving the students an opportunity to challenge each other. Group participation would also be appropriate in adding and subtracting in which students take turns presenting an addition or subtraction problem to their partner.

In learning to count, the teacher can encourage students to count things as they travel around, like the number of cars they pass going the opposite direction between home and the grocery store, or the number of concrete squares on the sidewalk in their walk home from school. How many outdoor light fixtures they see around the school. The number of trees in the city park. The possibilities for practicing counting are almost limitless. Students can be encouraged to find their own *things* to count.

In such counting exercises numbers will quickly become well known to the student. And this will bring comfort in dealing with numbers.

Additionally, there are, sprinkled throughout, simple drawings to color in, providing an activity separate from math and providing some fun in the learning experience.

TABLE OF CONTENTS

An Introduction to Numbers — 1	Adding Practice — 21
ZERO — 2	More Adding Practice — 22
Decimal Number System — 3	Basic Addition — 23
Counting — 4	Linear Form Refresher — 24
Digits — 5	Linear Addition Practice — 25
Object Counting — 6	More Linear Addition Practice — 26
Our First Two-Digit Number — 7	More on the Equal Sign (=) — 27
Two Digit Number Counting — 8	Tools — 28
The Twenties and After — 9	Subtraction - The Next Tool — 29
Counting by TWO — 10	Number Line Subtraction — 30
Counting by FIVE — 11	Number Line Subtraction - A Different View — 31
Counting by TEN — 12	Subtraction by Object Count — 32
Counting by TWENTY — 13	Adding and Subtracting — 33
Counting by TWENTY-FIVE — 14	Subtraction Table [Explained] — 34
Understanding Counting — 15	Subtraction Table — 35
Adding Numbers — 16	Subtraction Practice [Page 1] — 36
Adding Numbers (Digits) [Visual Presentation] — 17	Subtraction Practice [Page 2] — 37
Number Line Adding — 18	Subtraction Logic — 38
Adding Digits — 19	Linear Subtraction Practice — 39
Digit Addition Table — 20	Conclusion — 40

This page left blank intentionally and may be used for notes.

An Introduction to Numbers

Numbers are everywhere in our world, from finance and money to space travel, to biology, to communications. Numbers play a role in everything technical, whether it be chemistry, building a bridge, planet movement, flying, building a home, pouring a concrete patio, following a recipe, or understanding why our bicycle can stay up on two wheels. Numbers and mathematics is the language of everything technical. Therefore, the more we know about numbers, the more possibilities we have open to us.

We will begin with one of the most important numbers of all... ZERO (0). There was a time when number systems did not have a ZERO (0) in them, and this caused considerable problems. For example, the Roman Numeral system did not have a ZERO (0). And although Roman Numerals still have their uses, such as identifying movie dates, doing arithmetic with Roman Numerals is very difficult.

Our number system has Arabic origins and there are ten distinct and different characters. They are ZERO to NINE, shown here.

0 1 2 3 4 5 6 7 8 9

Can you think of things that require numbers to do?

For now, these ten characters can make any number we will need.

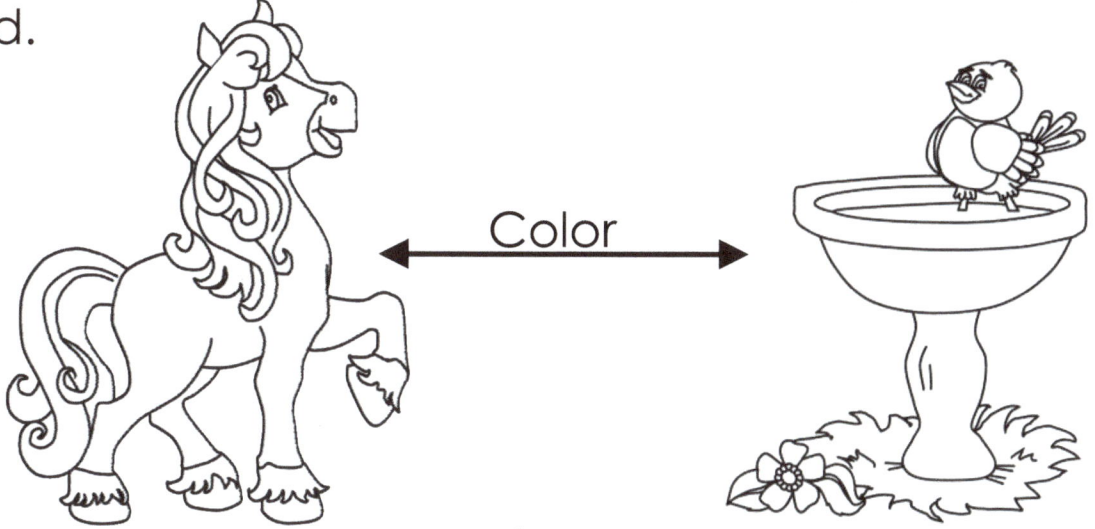

Color

ZERO - 0

The number zero (0) is an empty value. It means that nothing being measured exists. If we are counting apples, and we have zero (0) apples, then we have no apples. If we are counting beans, and we have zero (0) beans, then we have no beans. If we are counting tens, and we have zero (0) tens (10), then we have no tens. If we are counting hundreds (100) and we have zero (0) hundreds, then we have no hundreds.

All integer numbers, both positive numbers and negative numbers, live on a number line. And zero (0) is right in the middle of all the numbers. It is neither positive or negative, therefore it has no sign (+ or -). To the right of zero (0) are the positive numbers, and to the left of zero (0) are the negative numbers, shown in the second number line.

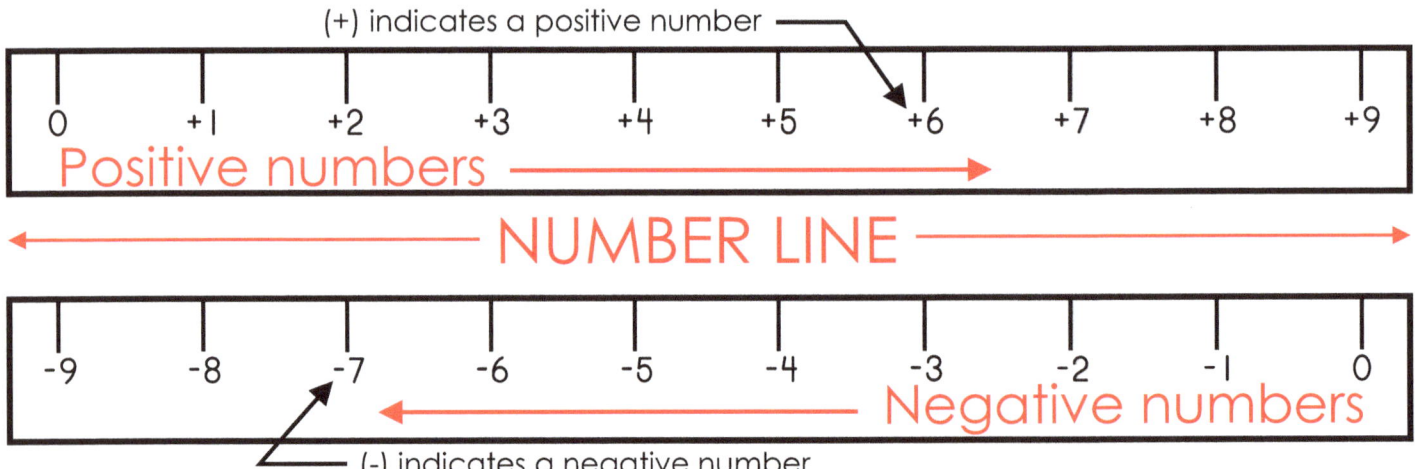

Showing in both number lines are the ten distinct characters which make up our number system. Every integer number in our number system can be represented with these ten characters. Integer numbers are whole numbers like, 1 - 2 - 3 - 4 - 5 and so on. We will learn more about other numbers in another workbook, for now we will just work with integers. And although plus (+) signs have been included on the number line, it is not common practice to put a plus (+) sign in front of a positive number. No sign means that the number is positive.

DECIMAL NUMBER SYSTEM

Our number system is called <u>decimal</u>. It is a base ten (10) number system and that is why there are ten distinct characters, zero (0) through nine (9).

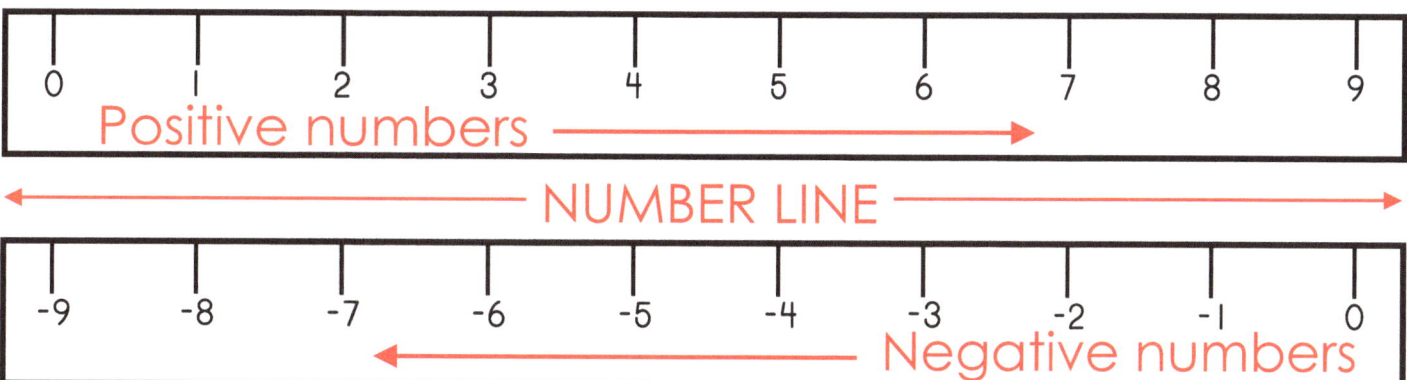

Our number lines above show the numbers zero (0) through nine (9), both positive and negative, but both number lines go on forever. We call this infinity (∞). Therefore, there is no end to the number line in either the positive, or negative direction. And although the number line is shown split as two different lines above, it is only one line, not two. There is no end to the numbers on the line, but that doesn't cause us any real concern, we will just use the numbers we need.

Numbers can represent anything we want. The number of cows in the pasture, the number of fence posts we need to put up a fence, the number of pounds of nails we need to build our house, the number of cubic yards of cement we need to build our patio, the number of cups of milk we need to add to our recipe, the thrust we need to get our space ship launched, the speed necessary to maintain an orbit at a specific altitude. Numbers are everywhere and the more we know about numbers, the more opportunities we will have to do fun and interesting things. The less we know about numbers, the more limited our opportunities will be.

COUNTING

Here we begin to learn to count. Counting generally involves identifying the number of a thing, like beans (🟡) or balls (🟢). OK, let's count and repeat as many times as needed.

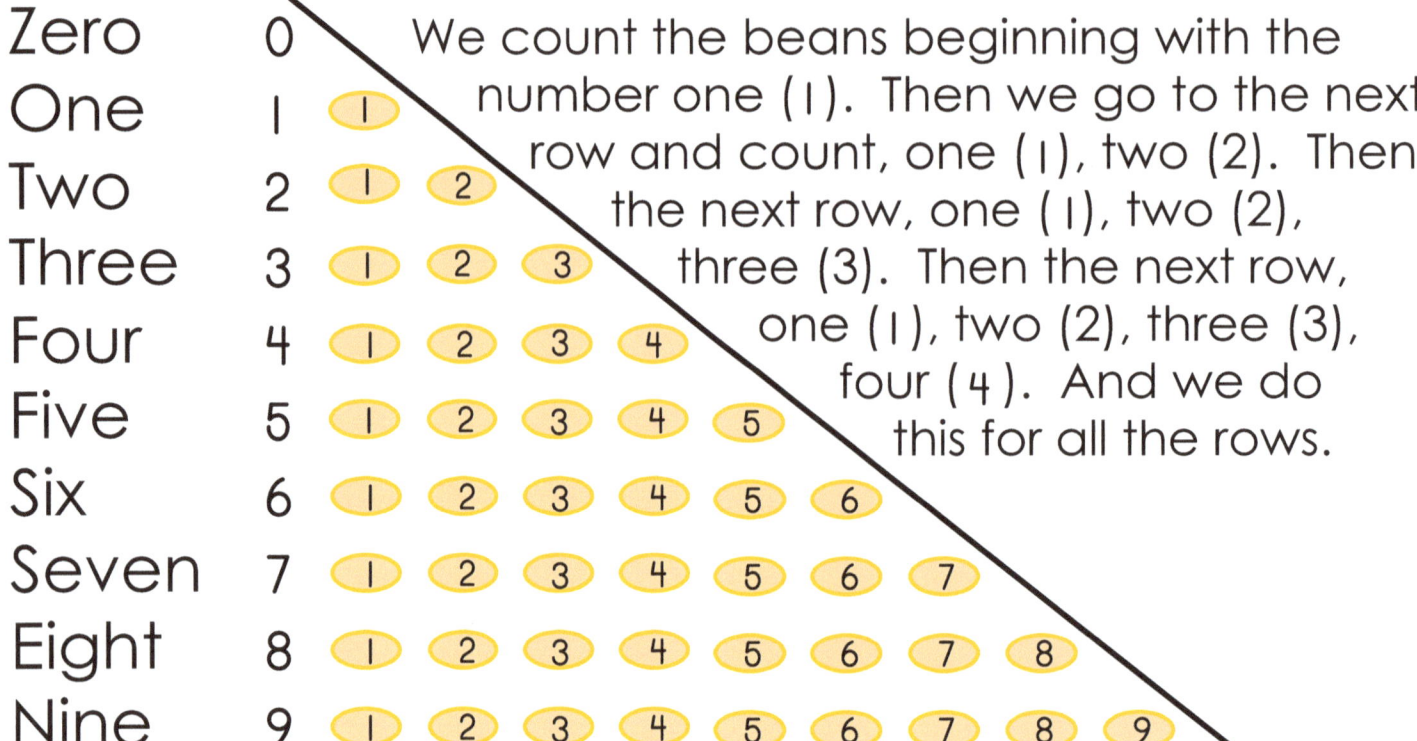

Zero 0
One 1
Two 2
Three 3
Four 4
Five 5
Six 6
Seven 7
Eight 8
Nine 9

We count the beans beginning with the number one (1). Then we go to the next row and count, one (1), two (2). Then the next row, one (1), two (2), three (3). Then the next row, one (1), two (2), three (3), four (4). And we do this for all the rows.

Count from zero (0) to nine (9) as many times as needed to be able to count without looking. When you can close your eyes and count zero (0) to (9) then go to the next page.

If there are other students, then form groups of two, three or four students. The first student starts the count with zero (0), and each student in turn advances the count until nine (9) is reached. If the group can count higher than nine, they should do so. Count as high as you can. If someone makes a mistake, start over. Keep counting until no mistakes are made. Make counting a game.   Color when permitted

DIGITS
0 1 2 3 4 5 6 7 8 9

Every number in our decimal number system is made up of these ten (10) characters or symbols that we call "digits." The term "digit" is Latin for finger, and we have ten fingers (count them), thus we have ten "digits," both on our hands and in our number system.

When we first learn to count, and then to add and subtract, and later on, multiply and divide, we find ourselves counting on our fingers. This is a normal and natural way in which we learn counting, and the beginning of arithmetic. "Arithmetic" is an elementary or very simple form of mathematics, it is where we begin to understand numbers and their uses.

Arithmetic is used by almost everyone, the butcher, the baker, the candle stick maker. It is used for tasks ranging from simple day-to-day counting to advanced science and business calculations. Arithmetic involves the study of combining numbers. In this book we will think of arithmetic as referring to the simpler operations of counting, addition, and subtraction of small number values, saving multiplication and division for a more advanced book.

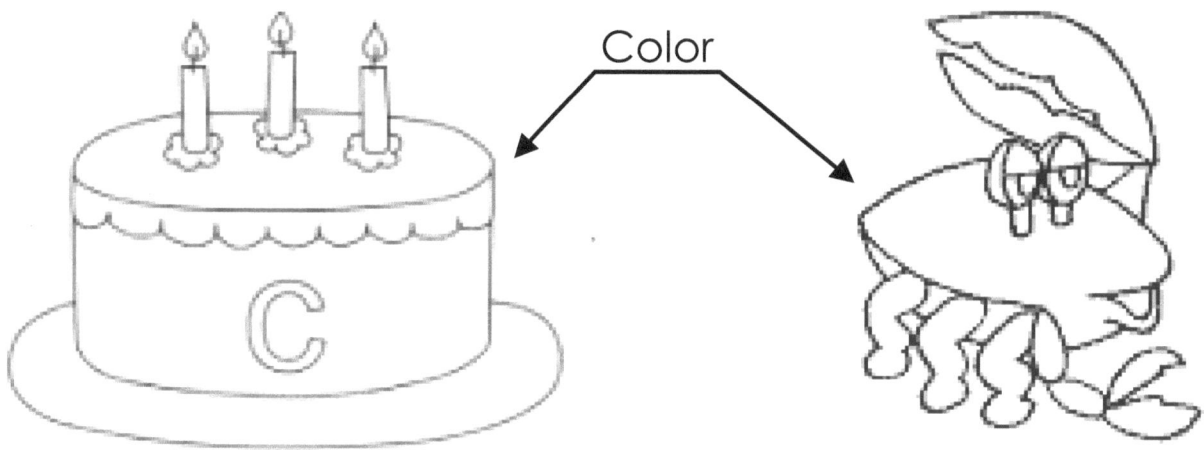
Color

OBJECT COUNTING

In this exercise, you will count objects and write the number of objects in the box. For example:

Butterflies: 6

Now you do it. Count the objects and write the number in the box.

Balloons:

Pansies:

Horses:

Bugs: Ants:

Tennis Balls:

Sweet Williams:

Badminton Birds:

On this page we have learned that numbers can represent all kinds of things. The things that numbers can represent is almost unlimited. Almost nothing can be made, measured or constructed without numbers, whether a dress, a house, a bridge, an office building, a recipe for stew, a pair of shoes, a football field, speed of a pitch, the course of a ship or the temperature outside. Numbers are everywhere.

OUR FIRST TWO DIGIT NUMBER

Counting zero (0) to nine (9) is a good skill, but we will need to count higher. Once we go above nine (9), we will have two digit numbers. Let's get started:

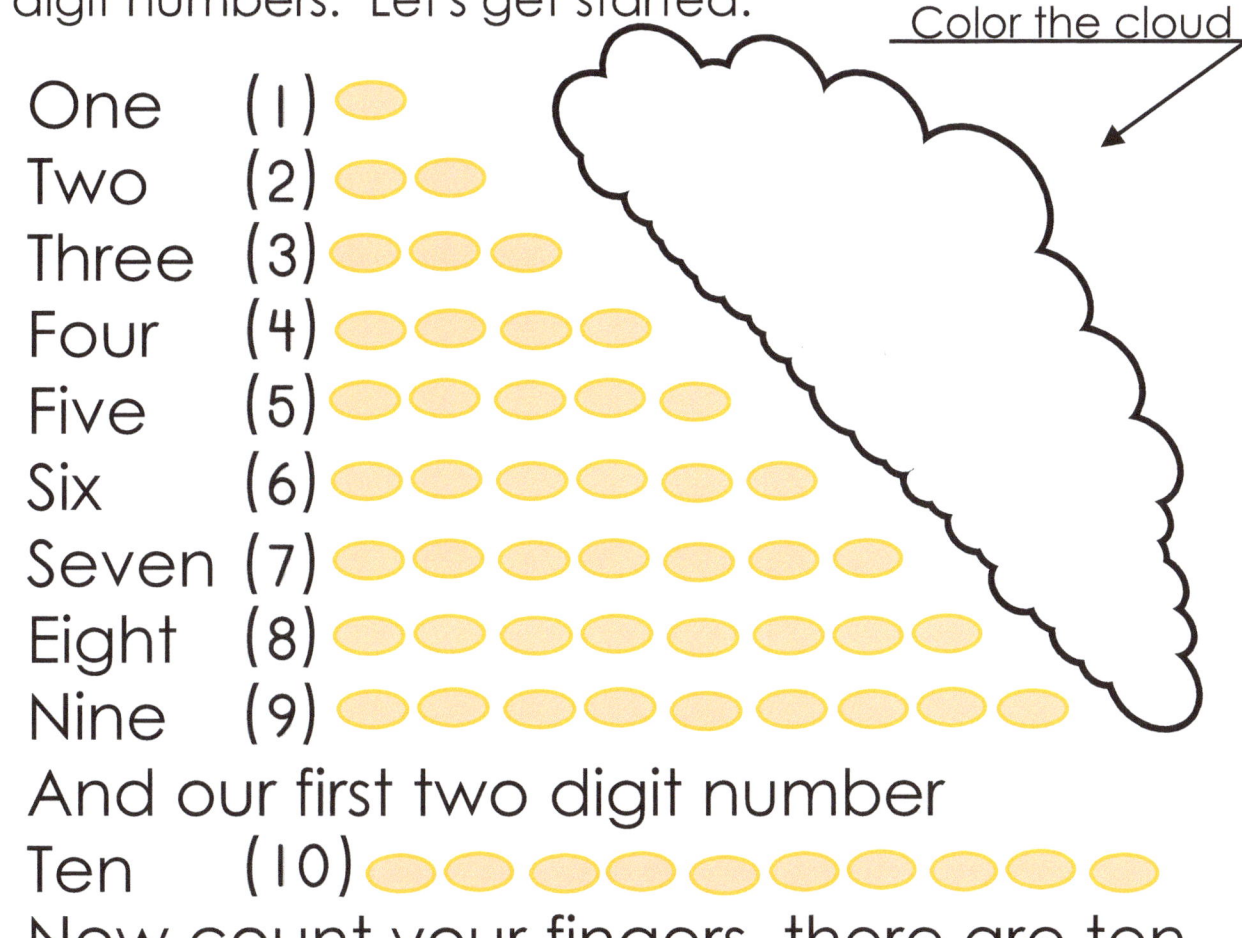

Color the cloud

One (1)
Two (2)
Three (3)
Four (4)
Five (5)
Six (6)
Seven (7)
Eight (8)
Nine (9)

And our first two digit number

Ten (10)

Now count your fingers, there are ten.

Color the picture

TWO DIGIT NUMBER COUNTING

Counting Higher Count the beans for each number

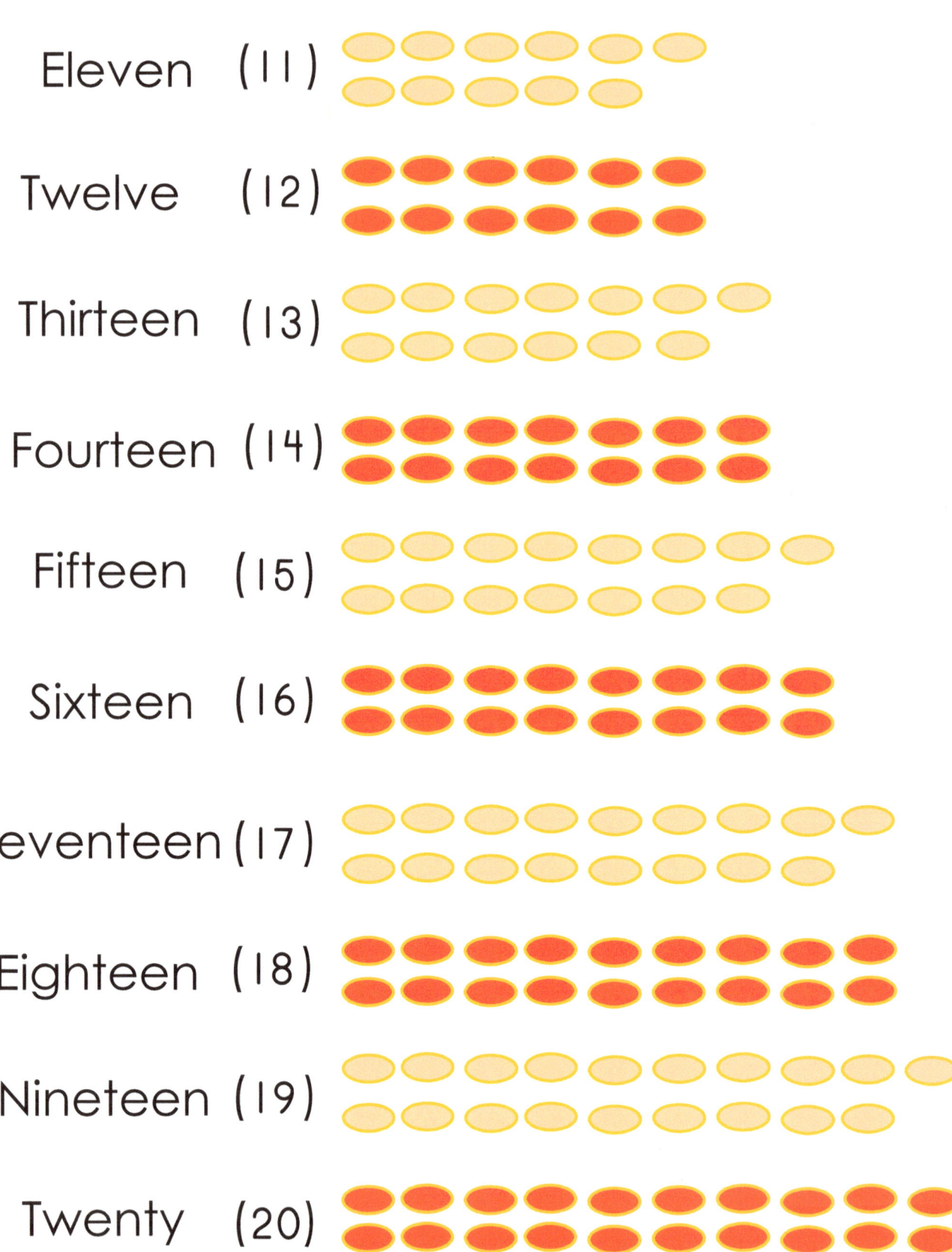

THE TWENTIES AND AFTER

When we get to the Twenties something interesting happens. The numbers are grouped by tens. In this I mean that we have the twenties followed by the thirties, forties, fifties, sixties, seventies, eighties, nineties and then begins the hundreds, a three digit number. And so, beginning at twenty, we count like this:

Twenty-one	(21)	Thirty-one	(31)
Twenty-two	(22)	Thirty-two	(32)
Twenty-three	(23)	Thirty-three	(33)
Twenty-four	(24)	Thirty-four	(34)
Twenty-five	(25)	Thirty-five	(35)
Twenty-six	(26)	Thirty-six	(36)
Twenty-seven	(27)	Thirty-seven	(37)
Twenty-eight	(28)	Thirty-eight	(38)
Twenty-nine	(29)	Thirty-nine	(39)
And then comes thirty...	(30)	And then comes forty...	(40)

If you can to a hundred (100), then color or draw below.

Draw your own picture

Color

COUNTING BY TWO

Thus far, we have counted by one. One, two, three, four, and so on. But it is quite beneficial to also count by other values, for example two (2). If we count by two, our starting value will generally be zero (0) and our first count will be two. And so we count by two:

2 - 4 - 6 - 8 - 10 - 12 - 14 - 16 - 18 - 20 - 22 - 24 - 26 - 28 - 30 - ...

This symbol means to continue as indicated

These are called <u>EVEN</u> numbers.

Now it's time to practice counting by two. If there are other students present, it may be beneficial to form groups of three, or two, and allow each student in turn, to advance the count by two. The first student would say, "two," and the next student would say, "four," and then the next student would say, "six."

Counting by two (2), fill in the four Number Lines below.

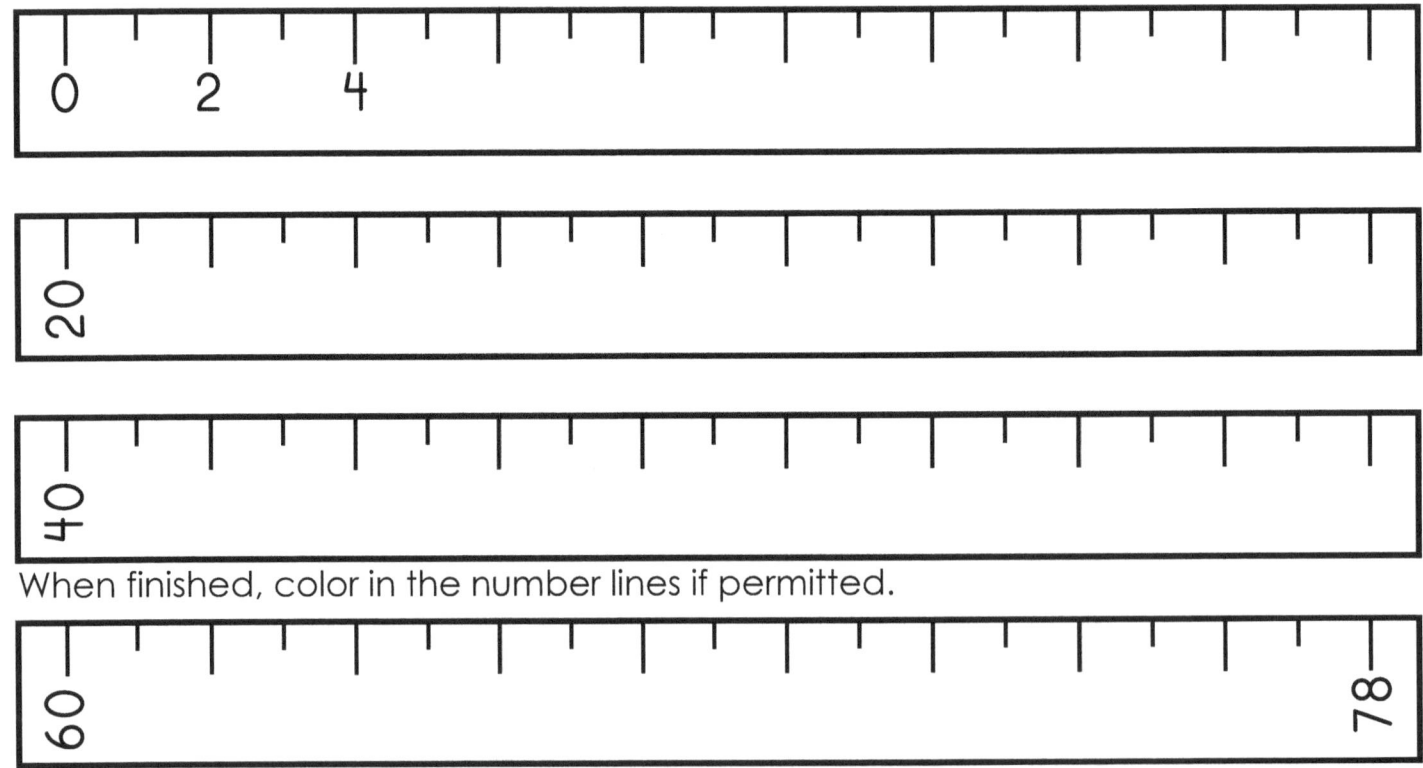

When finished, color in the number lines if permitted.

COUNTING BY FIVE

Another value by which to count is five (5). Counting by five is much the same as counting by two, except that the count advances by five rather than two. As before, our starting value is zero (0) and five (5) becomes the first value in our count.

$$5 - 10 - 15 - 20 - 25 - 30 - 35 - ...$$

This symbol means to continue as indicated

Once again it's time to practice counting, this time by five. And as before, if there are other students it may be beneficial to form groups of three, or two, and allow each student in turn, to advance the count by five. The first student would say, "five," and the next student would say, "ten," and then the next student would say, "fifteen."

Counting by five (5), fill in the three Number Lines below.

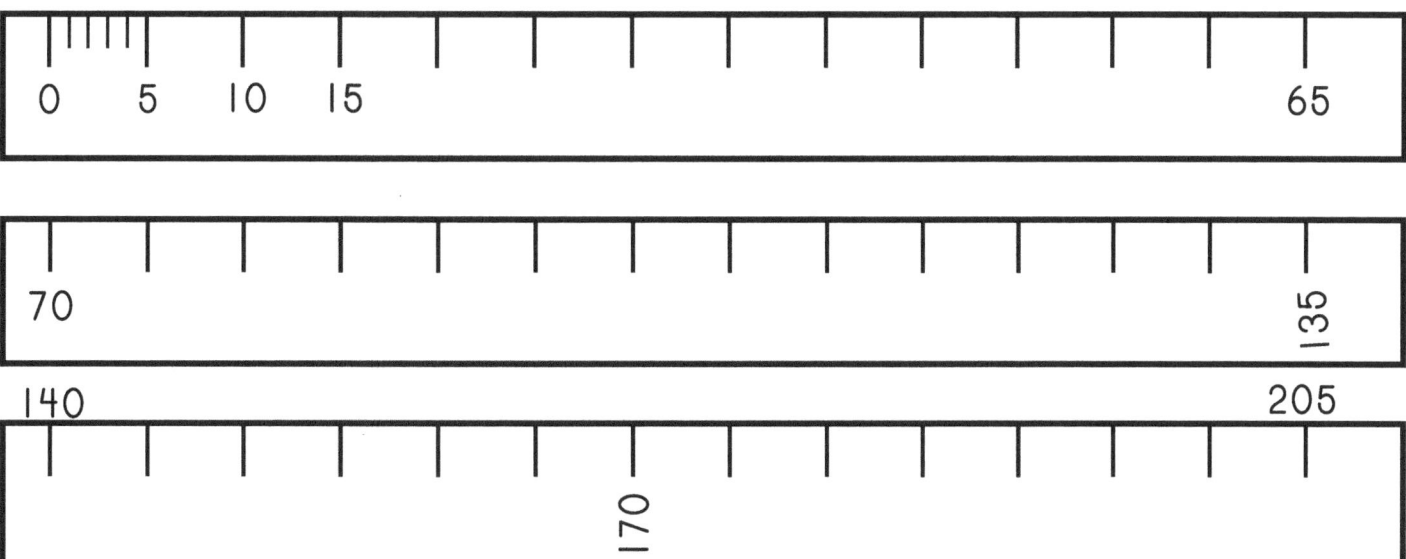

In reality, the number lines above are one number line, there is just not enough room on the paper to show them as one number line. The number line has all numbers on it. Positive numbers to the right of zero (0), and negative numbers to the left of zero (0). Since these numbers are to the right of zero (0), they are all positive.

COUNTING BY TEN

The value we will learn to count by this time, is ten (10). Counting by ten is much the same as counting by two and five, except that the count advances by ten rather than two or five. Our starting value is zero (0) and ten (10) becomes the first value in our count.

10 - 20 - 30 - 40 - 50 - 60 - 70 - ...

This symbol means to continue as indicated

How high can you count?

OK, let's practice counting by ten. Ten, twenty, thirty... And as before, if there are other students, it may be beneficial to form groups of three, or two, and allow each student in turn, to advance the count by ten. The first student would say, "ten," and the next student would say, "twenty," and then the next student would say, "thirty."

Counting by ten (10), fill in the next three Number Lines.

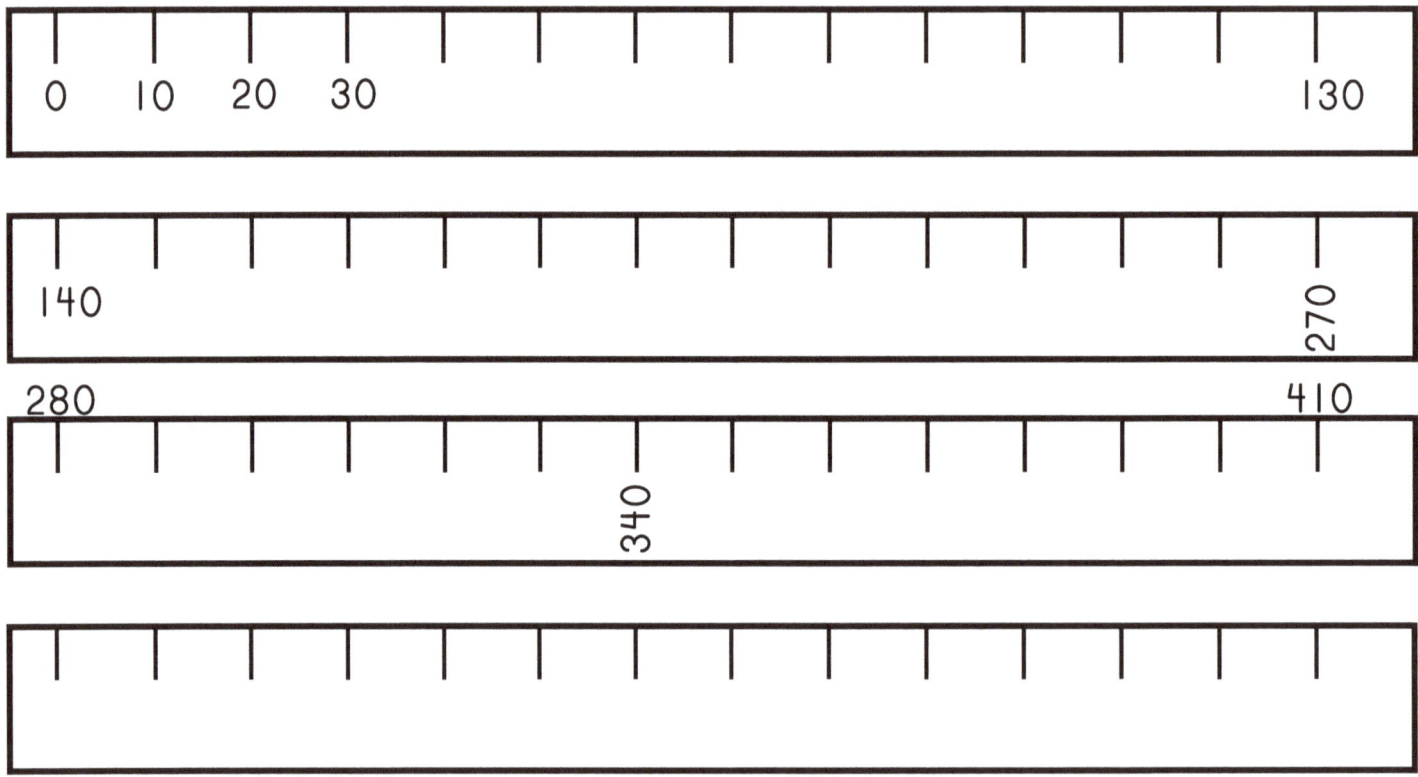

Bonus number line

COUNTING BY TWENTY

The value we will learn to count by this time, is twenty (20). Counting by twenty (20) is very closely related to counting by two, except that instead of counting 2 - 4 - 6 - 8 - 10 - ..., we count 20 - 40 - 60 - 80 - 100 - ... The count advances by twenty rather than two. Our starting value is zero (0) and twenty (20) becomes the first value in our count.

20 - 40 - 60 - 80 - 100 - 120 - 140 - ...

This symbol means to continue as indicated

How high can you count?

OK, let's practice counting by twenty (20). Twenty, forty, sixty... And as before, if there are other students, it may be beneficial to form groups of three, or two, and allow each student in turn, to advance the count by twenty. The first student would say, "twenty," and the next student would say, "forty," and then the next student would say, "sixty."

Counting by twenty (20), fill in the next three Number Lines.

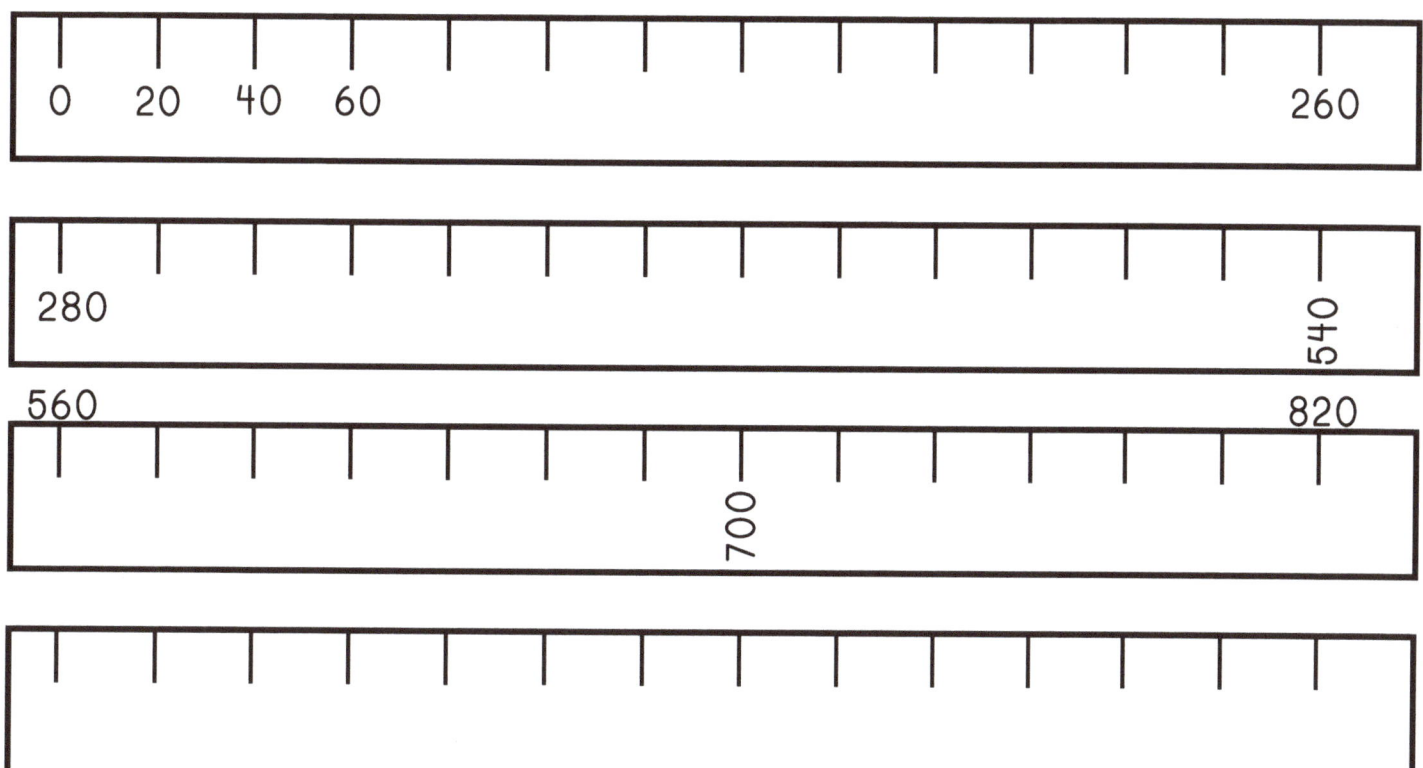

Bonus number line

COUNTING BY TWENTY-FIVE

The last value we will learn to count by is twenty-five (25). Counting by twenty-five is not any different than counting by two, five or ten except that the count advances by twenty-five rather than two, five or ten. Our starting value is still zero (0) and twenty-five (25) becomes the first value in our count.

25 - 50 - 75 - 100 - 125 - 150 - 175 ...   How high can you count?

This symbol means to continue as indicated ⬑

OK, let's practice counting by twenty-five. Twenty-five, fifty, seventy-five, a hundred, one-hundred-twenty-five, one-hundred-fifty... And as before, if there are other students, it may be beneficial to form groups of three, or two, and allow each student in turn, to advance the count by twenty-five. The first student would say, "twenty-five," and the next student would say, "fifty," and the third student would say, "seventy-five."

Counting by twenty-five (25), fill in the three Number Lines below.

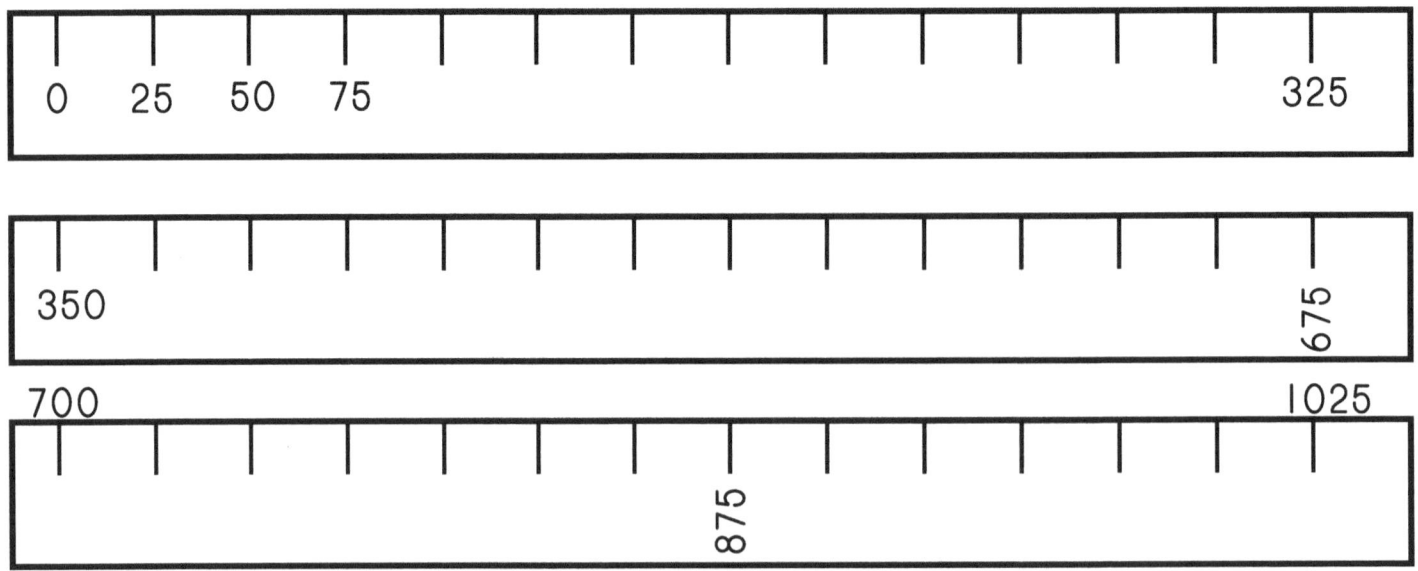

Remember, it's all one big number line.

UNDERSTANDING COUNTING

It is important to see that counting is based on the first ten digits, zero (0) to nine (9). Once we get to twenty, we say the group number, twenty, thirty, forty, and so on, followed by a digit number one (1) to nine (9). And it's like this all the way to the first three digit number, one-hundred (100). Then we have two-hundred (200), three-hundred (300) and so on. Here are some examples:

Sixty-seven .... (67)
Ninety-four .... (94)
One-hundred-four .... (104)
One-hundred-fifteen .... (115)
One-hundred-twenty-eight .. (128)
One-hundred-eighty-eight ... (188)
Two-hundred-fifty-three ...... (253)
Six-hundred-ninety-nine ...... (699)
Nine-hundred-forty-two ...... (942)

After the hundreds, which are three digit numbers, comes the thousands, a four digit number. As was pointed out earlier, the number system goes on forever, but we just work with the numbers we need.

Bonus: In the space below, count by three, beginning at zero (0). Your first count will be three (3), followed by six (6), nine (9), twelve (12), and so on. How high can you count by three (3)?

ADDING NUMBERS

We may not have recognized it, but we've been adding numbers already. When we count by one, we are simply adding one to the count. In adding numbers we will encounter two new symbols, they are the plus sign (+), which tells us to add or combine the numbers, and the equal sign(=). In general, what is on the left side of the equal sign has the same value as what is on the right side of the equal sign. Here are simple examples:

```
0 + 1 = 1     Zero plus one equals one.
1 + 1 = 2     One plus one equals two.
2 + 1 = 3     Two plus one equals three.
3 + 1 = 4     Three plus one equals four.
4 + 1 = 5     Four plus one equals five.
5 + 1 = 6     Five plus one equals six.
6 + 1 = 7     Six plus one equals seven.
7 + 1 = 8     Seven plus one equals eight.
8 + 1 = 9     Eight plus one equals nine.
9 + 1 = 10    Nine plus one equals ten.
```

Counting by ones is the same as adding one on each count.

10 + 1 = 11 ← Ten plus one equals eleven.
Now you try it.

11 + 1 =	16 + 1 =	21 + 1 =
12 + 1 =	17 + 1 =	22 + 1 =
13 + 1 =	18 + 1 =	23 + 1 =
14 + 1 =	19 + 1 =	24 + 1 =
15 + 1 =	20 + 1 =	25 + 1 =

The left side of the equal sign has the same value as the right side.

ADDING NUMBERS (digits)

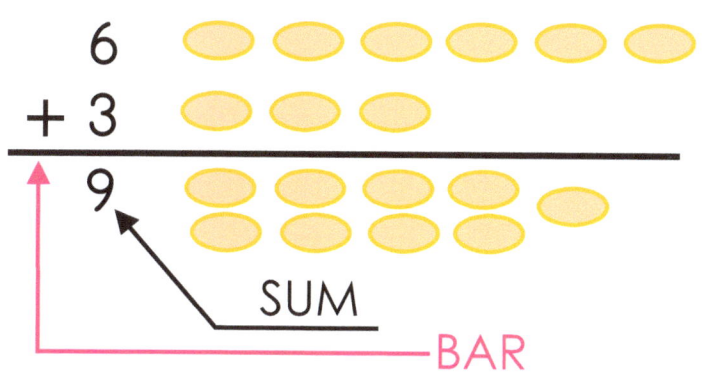

In the world of numbers we will find need to add numbers larger than one to the current value. For example, suppose we have six beans, and we add three more beans to the six we already have. How many beans will we have after we have added three more? As you can see, we can count the beans, and that may be easier now, but later, counting will be very difficult when numbers get larger. Therefore we must learn to add numbers and get the sum, which we might think of as the answer.

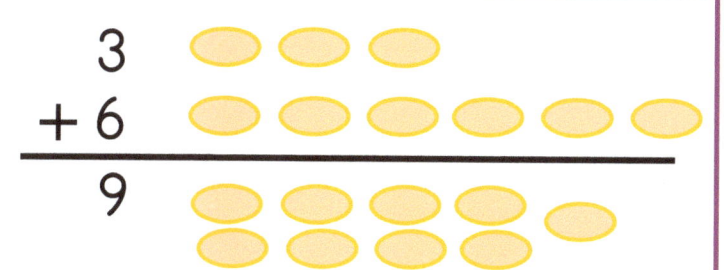

As you can see, it does not matter if we add three (3) to six (6), or if we add six (6) to three (3), the sum is the same.

Adding, as presented here, is a process of combining one positive number with another positive number to find the quantity or sum of those positive numbers. If you count the beans above and below the bar, there are the same number of beans.

Practice: Starting at zero (0), count to a hundred (100) by five (5).

NUMBER LINE ADDING

On the previous page we added three (3) beans to six (6) beans. In the beginning stages of learning to add, it sometimes helps to see the process of addition taking place on a number line. Below we see six (6) beans represented by the blue bar, and three (3) beans represented by the red bar. Now we are going to add them together.

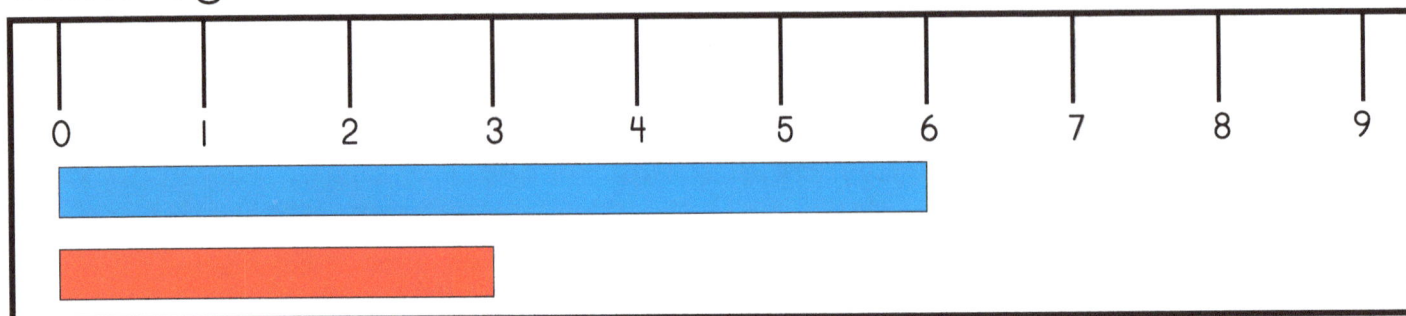

As we can see below, when we add three (3) to six (6), the sum (answer) is nine (9).

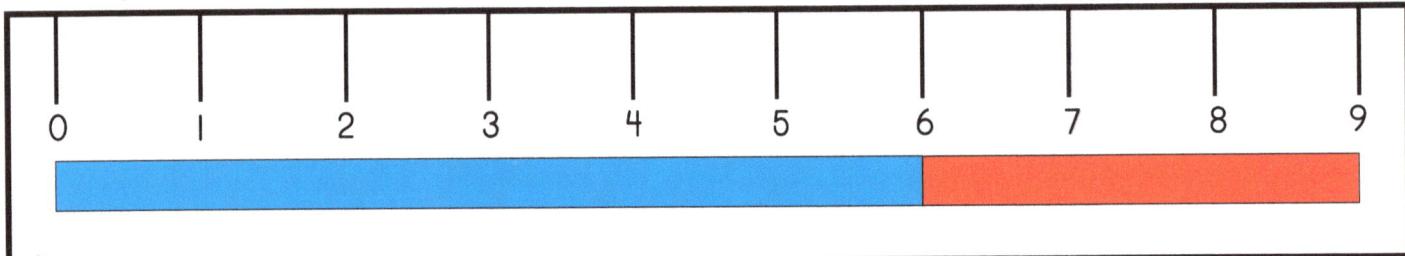

While we call this adding, it is actually a process of combining two positive numbers. The two number lines above provide a graphic or visual presentation of this process of combining. The student should probably think in terms of combining numbers and this will help in comprehension of other mathematical processes he, or she, will encounter in the future. For example, it is possible to add a negative number to a positive number, and calling such a process *adding* can be a little confusing. But thinking of the process as combining a positive number with a negative number can remove confusion.

ADDING DIGITS

In the world of adding (combining) numbers, our ability to add (combine) depends primarily on our ability to add (combine) the digits zero (0) to nine (9) quickly in our head. Therefore we must learn to quickly calculate the sum of all two number combinations of the first ten digits. The following page contains a TABLE to aid the student in learning to add (combine) all combinations of the first ten digits.

To use the table, put a left hand finger on a number in the left column, and then put a right hand finger on a number in the top row to add (combine) to the number chosen in the left column. Now move your left hand finger to the right, and move your right hand finger down. Where they come together is the sum of adding (combining) the two numbers that have been selected. This table must be learned and committed to memory. Practice until you no longer need the table to add (combine) any two of the first ten digits. <u>Using your pencil, find a way through the maze.</u>

DIGIT ADDITION TABLE

+	0	1	2	3	4	5	6	7	8	9
0	0	1	2	3	4	5	6	7	8	9
1	1	2	3	4	5	6	7	8	9	10
2	2	3	4	5	6	7	8	9	10	11
3	3	4	5	6	7	8	9	10	11	12
4	4	5	6	7	8	9	10	11	12	13
5	5	6	7	8	9	10	11	12	13	14
6	6	7	8	9	10	11	12	13	14	15
7	7	8	9	10	11	12	13	14	15	16
8	8	9	10	11	12	13	14	15	16	17
9	9	10	11	12	13	14	15	16	17	18

ADDING PRACTICE

Below are problems to practice adding (combining positive numbers). Can you do it without going back and looking at the table? If you cannot then a little more practice is needed to learn the sum of all combinations of the first ten digits.

4	3	2	1	0	1	2
+9	+8	+7	+6	+5	+4	+3

4	2	3	4	1	2	3
+2	+4	+4	+3	+5	+2	+6

4	0	1	0	2	1	2
+4	+0	+3	+3	+6	+0	+9

3	0	2	0	4	2	1
+9	+2	+0	+4	+0	+5	+2

3	0	3	0	3	4	1
+2	+9	+0	+6	+3	+6	+9

1	4	3		4	3	4
+7	+7	+7		+5	+5	+8

MORE ADDING PRACTICE

Remember, we must be able to add (combine) all two digit combinations of the first ten digits, zero (0) to nine (9), from memory. The student must practice this addition (combining) until it is no longer necessary to look at the addition table on page 20.

```
   2      3      4      4      1      3      3      4
 + 1    + 1    + 1    + 8    + 8    + 7    + 5    + 5
 ───    ───    ───    ───    ───    ───    ───    ───


   9      8      7      6      5      5      6      7
 + 0    + 9    + 0    + 9    + 0    + 4    + 4    + 3
 ───    ───    ───    ───    ───    ───    ───    ───


   5      6      7      6      8      9      6      7
 + 7    + 6    + 5    + 0    + 4    + 3    + 5    + 8
 ───    ───    ───    ───    ───    ───    ───    ───


   8      2      4      8      9      9
 + 8    + 8    + 7    + 2    + 1    + 5
 ───    ───    ───    ───    ───    ───
```

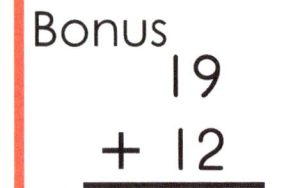

Bonus
  19
+ 12
────

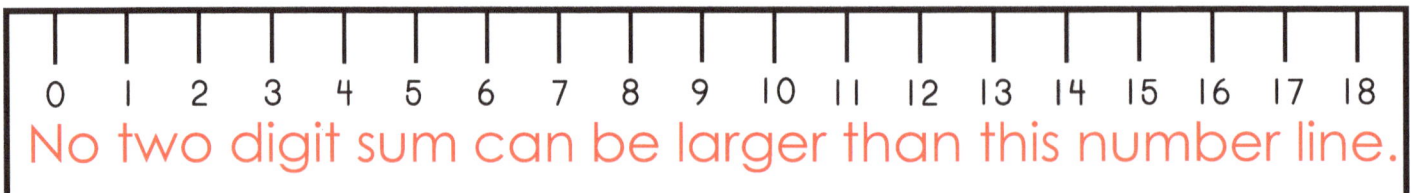

No two digit sum can be larger than this number line.

BASIC ADDITION

Calculations in mathematics are very dependent on addition (combining positive numbers). Remembering the sum of all two digit number combinations will serve the student all his life. Practice until you can add (combine) all two digit combinations.

```
   7      5      9      8      5      6      9
  +4     +1     +7     +0     +5     +2     +4
  ___    ___    ___    ___    ___    ___    ___

   8      5      7      8      9      8      5
  +6     +3     +1     +3     +8     +1     +2
  ___    ___    ___    ___    ___    ___    ___

   7      7      9      5      6      8      5
  +6     +7     +9     +6     +3     +5     +9
  ___    ___    ___    ___    ___    ___    ___

   6      5      7      6      8      9      6
  +1     +8     +9     +7     +7     +2     +8
  ___    ___    ___    ___    ___    ___    ___
```

Bonus
```
   9      7     17     11     13     15     18
  +6     +2    +15    +11    +14    +19    +17
  ___    ___   ___    ___    ___    ___    ___
```

| | | | | | | | | | | | | | | | | | | |
|0|1|2|3|4|5|6|7|8|9|10|11|12|13|14|15|16|17|18|

No two digit sum can be larger than this number line.

LINEAR FORM REFRESHER

Standard Form

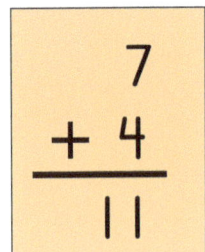

Thus far, addition problems have mostly taken on the form shown here. Although presented earlier, the linear form is repeated here. It is a common form with which the student should become familiar. The same problem in linear form appears in this manner:

Linear Form →

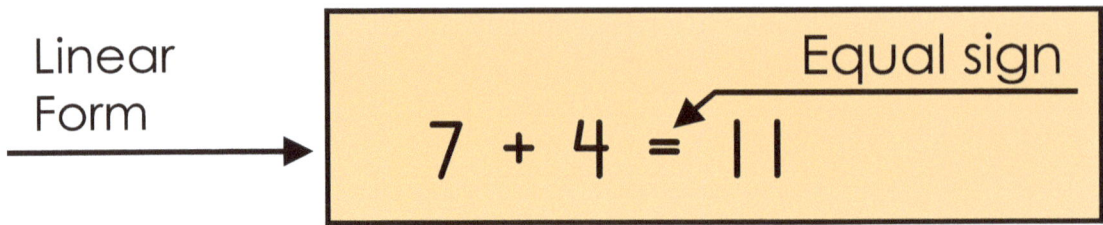

The equal sign (=) has a meaning in mathematics. In general it means that what is on the left side of the equal sign, has the same value as what is on the right side of the equal sign. As the student progresses further into mathematics this form of problem will occur more often, therefore, it is appropriate that this form be introduced at this time.

The following pages contain linear form practice problems for the student to complete.

Use your pencil and find your way through this maze

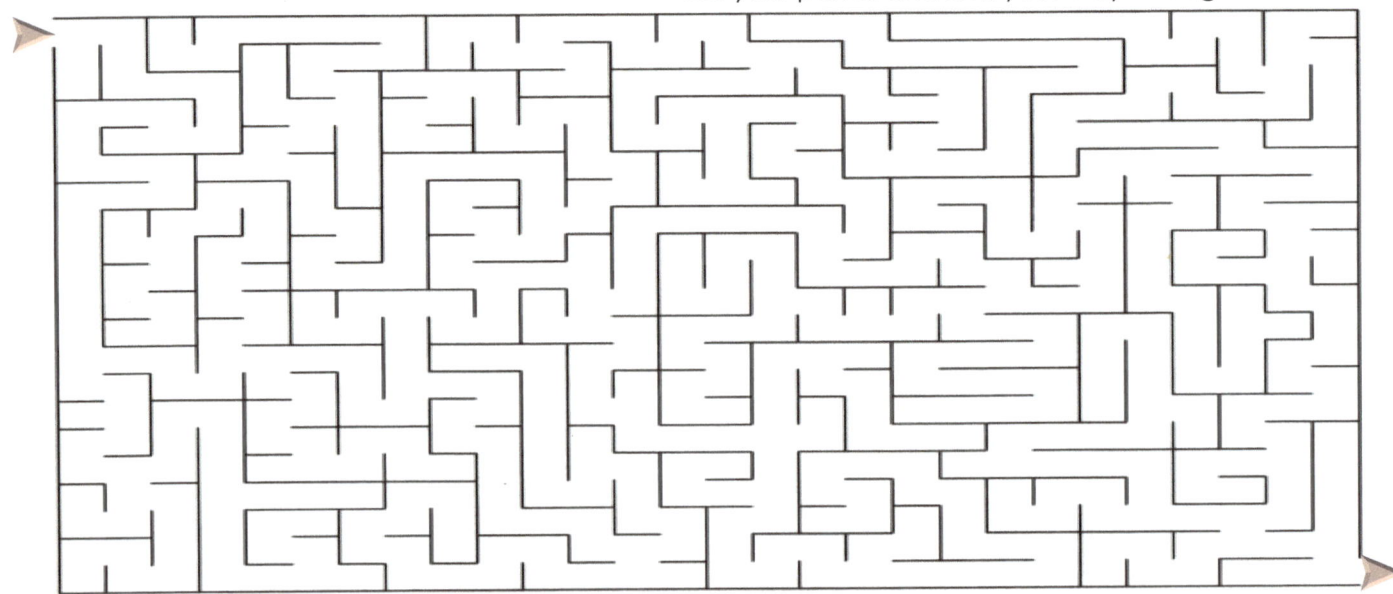

LINEAR ADDITION PRACTICE

You have now practiced adding two digit combinations of the first ten digits. Beginning on this page you will further your knowledge and skill at adding. The problems are presented in linear format as shown on the previous page. Let's begin:

4 + 9 =	0 + 4 =	1 + 1 =
3 + 8 =	1 + 8 =	0 + 1 =
2 + 6 =	3 + 9 =	2 + 3 =
1 + 4 =	3 + 0 =	3 + 7 =
0 + 2 =	0 + 9 =	0 + 3 =
1 + 0 =	2 + 9 =	4 + 7 =
2 + 2 =	1 + 5 =	2 + 8 =
3 + 6 =	0 + 0 =	0 + 7 =
4 + 6 =	1 + 6 =	2 + 4 =
0 + 6 =	4 + 4 =	4 + 5 =
3 + 3 =	2 + 0 =	2 + 1 =
4 + 0 =	4 + 2 =	0 + 5 =
3 + 2 =	4 + 1 =	2 + 5 =
3 + 4 =	3 + 5 =	4 + 3 =
0 + 8 =	1 + 3 =	3 + 1 =
1 + 9 =	1 + 7 =	1 + 2 =
Bonus Problems	16 + 3 =	12 + 7 =
15 + 7 =	16 + 6 =	19 + 3 =
13 + 7 =	4 + 16 =	5 + 15 =
19 + 6 =	7 + 18 =	7 + 17 =
9 + 16 + 7 =	7 + 7 + 7 =	7 + 17 + 19 =

MORE LINEAR ADDITION PRACTICE

This is the final page of linear addition problems. Are you able to answer the problems without looking back in the addition table? If not, just keep practicing.

7 + 0 =	8 + 2 =	4 + 8 =
2 + 7 =	7 + 4 =	6 + 5 =
9 + 4 =	7 + 7 =	6 + 9 =
5 + 9 =	7 + 9 =	6 + 4 =
6 + 0 =	5 + 1 =	9 + 7 =
6 + 1 =	8 + 3 =	5 + 0 =
7 + 5 =	9 + 2 =	8 + 4 =
7 + 8 =	9 + 0 =	5 + 2 =
5 + 5 =	6 + 7 =	7 + 1 =
9 + 1 =	5 + 8 =	6 + 3 =
8 + 1 =	9 + 3 =	5 + 3 =
7 + 3 =	5 + 6 =	6 + 8 =
8 + 9 =	8 + 8 =	9 + 8 =
6 + 2 =	5 + 7 =	8 + 7 =
5 + 4 =	8 + 6 =	7 + 2 =
8 + 0 =	9 + 5 =	6 + 6 =
8 + 5 =	9 + 6 =	7 + 6 =
9 + 9 =		

Bonus Problems

3 + 4 + 2 =	6 + 14 + 4 =	5 + 5 + 9 =
3 + 5 + 4 =	16 + 1 + 2 =	16 + 9 + 5 =
3 + 7 + 14 =	13 + 14 + 4 =	19 + 14 + 11 =
9 + 15 + 8 =	13 + 14 + 12 =	19 + 19 + 20 =

MORE ON THE EQUAL SIGN (=)

As was said before, the equal sign (=) means that the left side has the same value as the right side. There is a special name for a problem with an equal sign (=), it is called an <u>equation</u>. To illustrate this concept, look at the "equation" in the first sand box.

$$7 + 4 = 11$$

This means that 7 + 4 on the left side of the equal sign has the same value as eleven (11) on the right side. But the next two equations are just as valid as the one above.

$$7 + 4 = 5 + 6 \quad \text{OR} \quad 9 + 2 = 8 + 3$$

BECAUSE $\quad 11 = 11$

Both equations are true and correct, because 11 = 11. This is an important concept that should be remembered. The equal sign (=) means that the left has the same value as the right. Seven plus four (7 + 4) and five plus six (5 + 6) are just another way of expressing eleven (11). They are all the same.

$$2 + 2$$

Two plus two (2 + 2) is another way of expressing four (4). In the empty sand box, can you think of another two number addition problem to express the value of four (4)?

TOOLS

The learning of mathematics is the acquisition of tools, and this is true for much of education whether it be chemistry, physics, biology, electronics, reading or writing and composition. The school system does not teach you what to write, it gives you the tools to write. Education gives you the ability to sound out words, to spell words, to properly construct a sentence and use punctuation. But the words come from you. And if you have retained the [educational] tools you have been given, then your words will make sense when read by others.

The same is happening in this workbook. You are being given tools to solve mathematical problems. Thus far these are the tools you've been given.

> Counting by one
> Counting by two
> Counting by five
> Counting by ten
> Counting by twenty
> Counting by twenty-five
> Introduction to the digits in the decimal number system
> Adding all combinations of digits
> Introduction to an equation
> Equation addition

As you progress in the learning of mathematics, you will be given many tools. In order to be good at math, or physics, or biology, or chemistry, or reading, or writing and composition, you must remember and take the tools you are given with you from year to year and throughout life.

SUBTRACTION - The next tool

Subtraction is the opposite of addition. Where addition is a process of combining one positive value (number) with another positive value (number), subtraction is the taking away of one value (number) from another value (number). Subtraction involves a new symbol called the <u>minus sign</u> (−). Take a look at the problem in the box and you will see the minus sign (−) properly placed. This is a problem in which you have nine (9) and three (3) is taken away from nine, leaving six (6). Let's take a look at this on the number line.

$$\begin{array}{r} 9 \\ -3 \\ \hline 6 \end{array}$$

Number Line

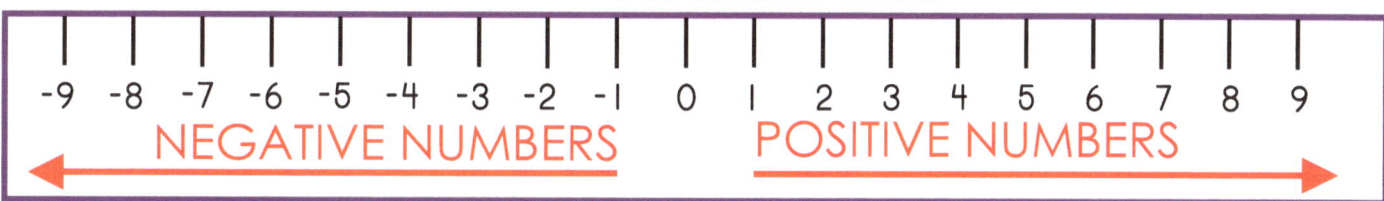

Because the number three (3) in the above problem has a negative sign to its left, it is considered a <u>negative number</u>. Both positive and negative numbers are on the number line. Positive numbers to the right of zero (0), negative numbers to the left. This is why it is more appropriate to think of both addition and subtraction as the combining of numbers. In addition, we are combining one positive number with another positive number. However, in subtraction we are combining a positive number with a negative number.

Suppose we were to subtract nine (9) from five (5)? Then what is the answer? Well, if we put our pencil on the number five (5), and count to the left nine numbers, our pencil will be on negative four (-4). This simple example shows why it is easier to think in terms of combining numbers rather than only adding and subtracting.

NUMBER LINE SUBTRACTION

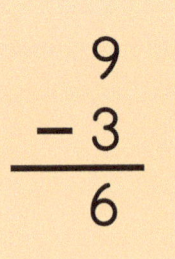

We bring our problem in the sand box forward to this page so that we can graphically see what is taking place in subtracting one number from another. Using the number line we will subtract three (3) from nine (9) to get our answer, six (6).

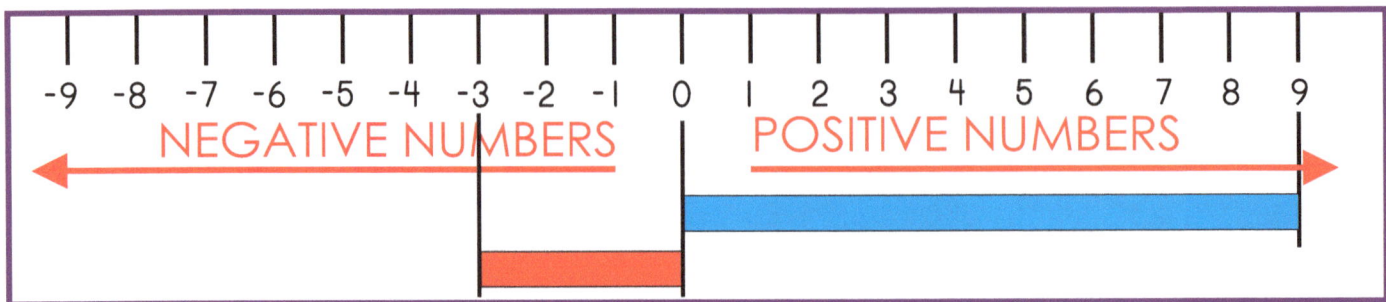

On the number line above, the blue line has a value of nine (9) and the red line has a value of negative three (-3). In the number line below, we combine (apply) our red negative three (-3) value to our blue positive nine (9) value and we can see that when we take three (3) away from nine (9) the answer is six (6).

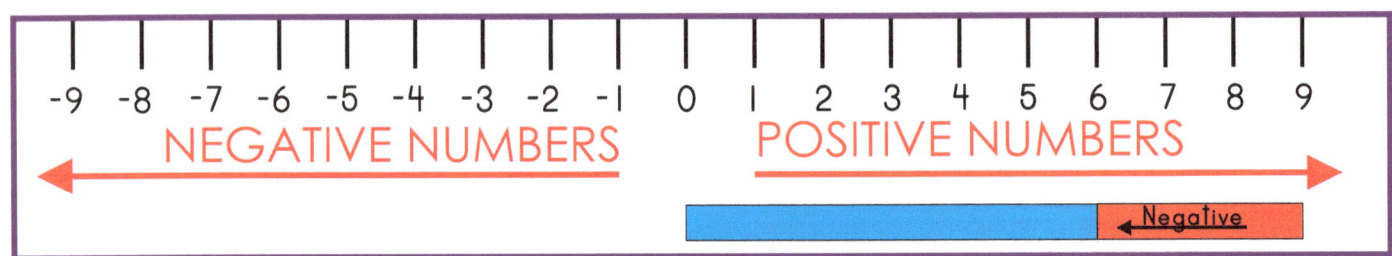

Count	Number
1	4
2	3
3	2
4	1
5	0
6	-1
7	-2
8	-3
9	-4

Suppose we were to take nine (9) from five (5). The process is the same. We put our pencil on five (5) and count nine (9) numbers to the left. The first number to the left is four (4), and we see that in the chart at the left. When we finish counting nine (9) numbers to the left of five (5), we find our answer is negative four (-4).

NUMBER LINE SUBTRACTION
A different view

Mathematics is flexible and can be viewed in a number of ways. Sometimes, viewing the same problem in a different way allows the student to see more clearly what is taking place. That is what we are going to do here, again with the number line. The quantity nine (9) is going to be presented twice on the number line, once to the right of zero (0), and again with three taken away by moving it three counts in the negative direction, left of zero (0).

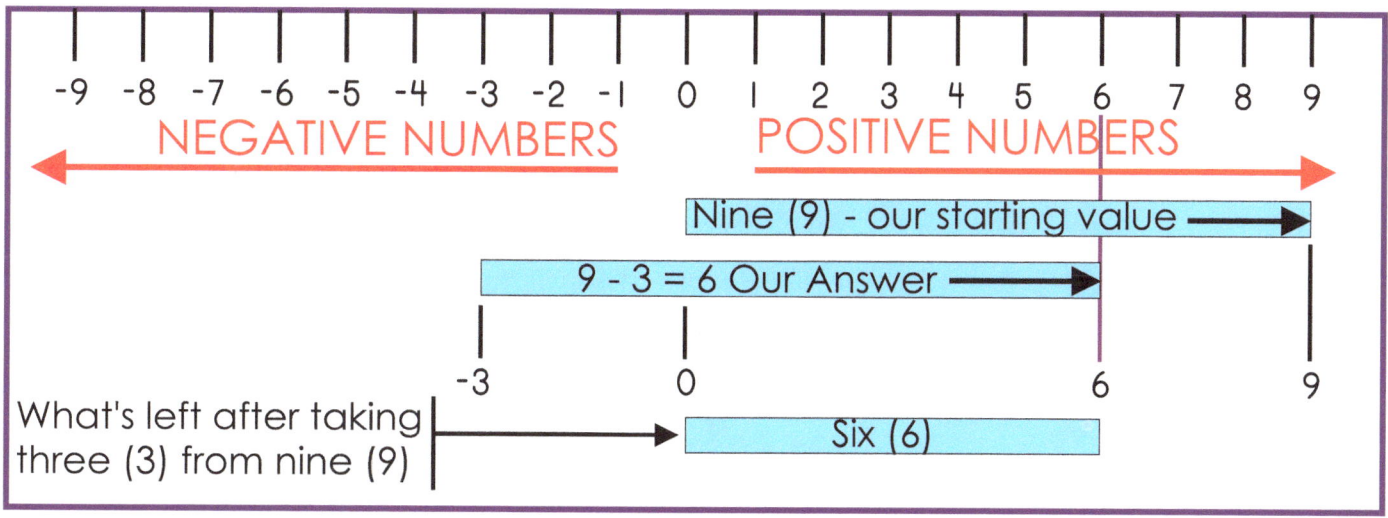

In the above number line example, all we have done is take the upper blue rectangle, representing a value of nine, dropped it down and slid it to the left of zero (0) three (3) counts. The line now has a positive value of six (6). This is subtraction, or more accurately, combining a positive and a negative number, in this case a positive nine (+9) with a negative three (-3), and six (6) is the result.

Extra Credit
Make a number line.

SUBTRACTION BY OBJECT COUNT

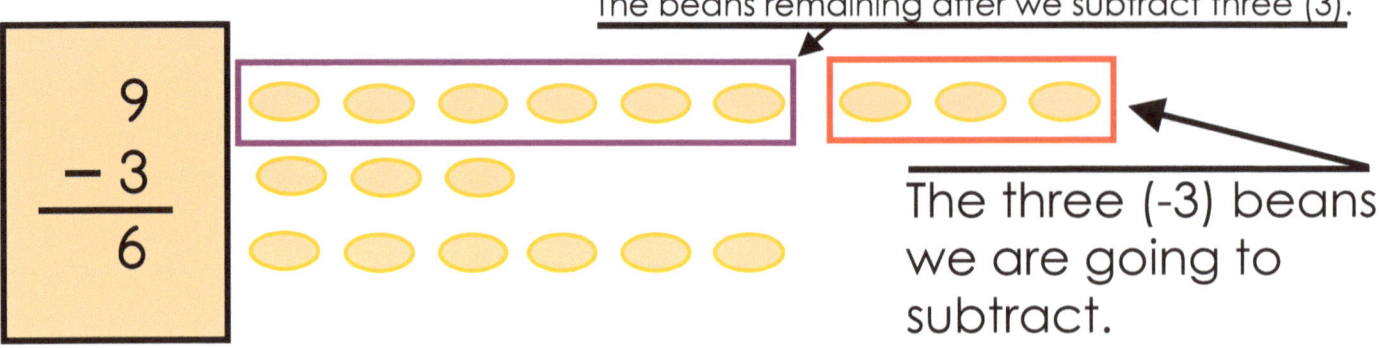

Nine (9) minus three (-3) equals six (6).

$$9 - 3 = 6$$

Once again we have brought our problem in the sand box forward. To the right of the number nine (9) there are nine (9) beans. Outlined in red are the three beans that we are going to subtract (take away) and they are shown to the right of the number three (3). What is left after we have taken three (3) beans from nine (9) beans is six (6) beans, shown to the right of the number six (6), our answer.

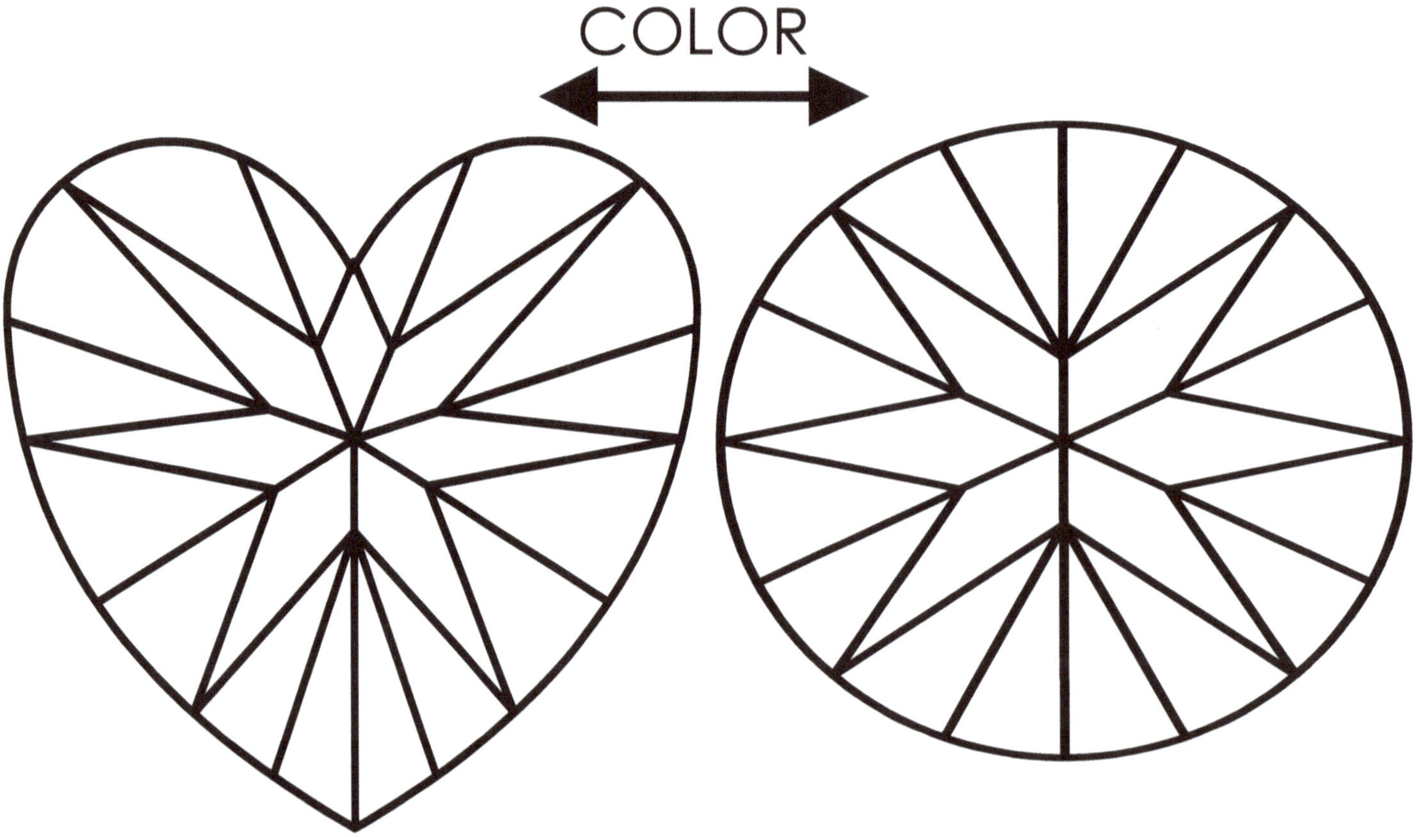

COLOR

ADDING AND SUBTRACTING

In adding and subtracting, we are combining numbers to get the answer (result). If we are to combine a blue positive five (+5) and a purple positive three (+3), we can see in our number line below that the result is a positive eight (+8). However, if we combine a blue positive five (+5) with a red negative three (-3), also seen in the number line below, the result is a positive two (2).

Addition is the process of combining one positive value with another postive value. And subtraction is the process of combining one positive value with one negative value. Therefore, addition and subtraction both involve the combining of two numbers to get an answer (result).

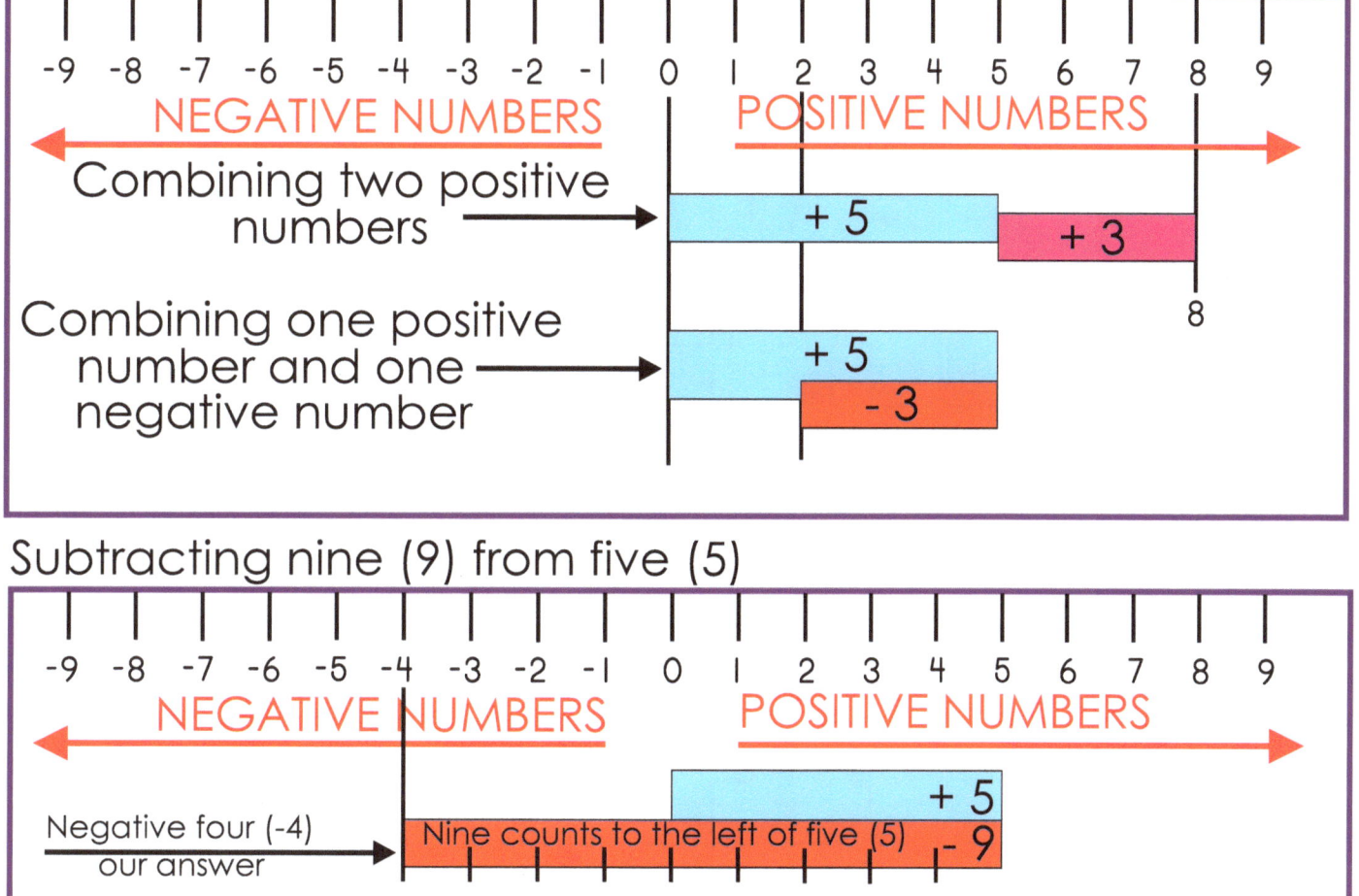

SUBTRACTION TABLE

On the next page is a basic subtraction table. It operates about the same as the addition table. The top row are the positive numbers and the left hand column contains the negative numbers the student wishes to subtract. As before, put a finger of the right hand on the positive number, and a finger of the left hand on the negative number to subtract. Move the right finger down, and the left finger to the right, and where the two fingers meet is the answer or result of the problem.

The table contains all combinations of two digit numbers that result in a positive or zero answer. A minus sign means that subtracting the number on the left from the number on the top results in a negative number answer. And right now we are not interested in negative number answers, that will come later in another book. For now, we are just interested in subtraction problems that present a zero or positive result or answer.

Can you find your way through the maze?

DIGIT SUBTRACTION TABLE

Bonus: Can you fill in the negative answers?

	0	1	2	3	4	5	6	7	8	9
0	0	1	2	3	4	5	6	7	8	9
-1	-	0	1	2	3	4	5	6	7	8
-2	-	-	0	1	2	3	4	5	6	7
-3	-	-	-	0	1	2	3	4	5	6
-4	-	-	-	-	0	1	2	3	4	5
-5	-	-	-	-	-	0	1	2	3	4
-6	-	-	-	-	-	-	0	1	2	3
-7	-	-	-	-	-	-	-	0	1	2
-8	-	-	-	-	-	-	-	-	0	1
-9	-	-	-	-	-	-	-	-	-	0

SUBTRACTION PRACTICE

On this page, we will practice subtraction of problems with two digits. We will limit our problems to only those that produce either a zero (0) or positive result.

3 − 2	2 − 2	9 − 7	4 − 2	6 − 1
7 − 5	8 − 0	9 − 9	2 − 0	5 − 2
6 − 4	8 − 4	0 − 0	7 − 1	5 − 5
9 − 2	5 − 0	7 − 3	6 − 6	1 − 0
5 − 4	3 − 3	8 − 5	2 − 1	9 − 1
4 − 4	6 − 0	8 − 8	9 − 6	3 − 0

The remainder of the problems are on the next page.

SUBTRACTION PRACTICE

On this page, we will practice subtraction of problems with two digits. We will limit our problems to only those that produce either a zero (0) or positive result.

1 − 1	9 − 5	8 − 2	6 − 5	5 − 3
3 − 1	9 − 0	6 − 3	7 − 6	4 − 3
8 − 3	7 − 7	5 − 1	9 − 3	8 − 7
7 − 4	6 − 2	7 − 0	8 − 6	4 − 0
9 − 4	7 − 2	8 − 1	9 − 8	4 − 1

As it was with addition, it will eventually be necessary to be able to subtract without using the subtraction table.

This concludes all combinations of subtraction involving the first ten digits that produce only a positive or zero result.

SUBTRACTION LOGIC

$$\begin{array}{r} 9 \\ -\ 4 \\ \hline \end{array}$$ Here we have the problem of subtracting four (4) from nine (9). To solve this problem it is only necessary to ask ourselves what number we add to four (4) to get nine. Since we already know our addition table, we know that we add five (5) to four (4) to get nine, therefore, our answer is five (5). Nine (9) minus four (4) equals five (5). Therefore:

$$\begin{array}{r} 9 \\ -\ 4 \\ \hline 5 \end{array}$$ **OR** $9 - 4 = 5$

The point here is this... If we learn one skill, in this case addition, it often will become a tool to use in development of another skill, in this case subtraction. Mathematics is like that as well as many other fields of learning, including physics, chemistry, biology, geology, electronics... virtually every technical field is a process of developing new skills, often based in using the skills we have already acquired. With new skills come new-thoughts, some of which will have never been thought of before. It is how we advance.

What makes learning other interesting things fun is having the tools with which to learn, and mathematics is a key tool. Mathematics is essentially the language of science. Mathematics is used to count your money and to put a space ship into orbit. Math is your means to open many doors to a lot of fun and interesting occupations and hobbies, such as planetary movement. All of learning is a choice, made by each student.

LINEAR SUBTRACTION PRACTICE

You have now practiced subtraction of two digit combinations of the first ten digits. As we did with addition, we shall now do with subtraction. Below are subtraction problems presented in linear format. Let's begin:

5 - 3 =	9 - 5 =	8 - 4 =
5 - 5 =	6 - 4 =	6 - 3 =
7 - 1 =	9 - 4 =	2 - 1 =
7 - 6 =	7 - 2 =	8 - 5 =
9 - 6 =	6 - 6 =	6 - 2 =
3 - 2 =	2 - 0 =	8 - 2 =
7 - 5 =	6 - 0 =	8 - 6 =
5 - 4 =	4 - 2 =	1 - 1 =
3 - 3 =	7 - 3 =	8 - 7 =
5 - 0 =	9 - 8 =	4 - 1 =
8 - 0 =	9 - 7 =	5 - 1 =
2 - 2 =	4 - 3 =	9 - 0 =
6 - 5 =	9 - 1 =	8 - 1 =
8 - 3 =	7 - 4 =	6 - 1 =
9 - 2 =	7 - 7 =	5 - 2 =
3 - 1 =	9 - 9 =	4 - 4 =
8 - 8 =	7 - 0 =	9 - 3 =

Just like sewing or repairing an engine, you must have a collection of the right tools for the job. However, there is another job that doesn't appear as such, it is living in our society. In our society, we learn to speak, to read, to write and, of course, mathematics, these are basic tools we are given to use as we live our lives. Our command of these tools will determine, to a great extent, how successful, and happy, our lives will be.

CONCLUSION

In this workbook the student has learned how to count by one (1), count by two (2), count by five (5), count by ten (10), count by twenty and count by twenty-five (25), add combinations of the first ten digits, zero (0) through nine (9), and subtract combinations of the first ten digits, zero (0) through nine (9) which result in a positive answer. And although there are negative answers and negative numbers. Study of this concept is reserved for a more advanced workbook.

Feedback may be directed to mhkeehn@gmail.com

Not all things are as they seem. Look at the objects below and consider them, they are an optical illusion.

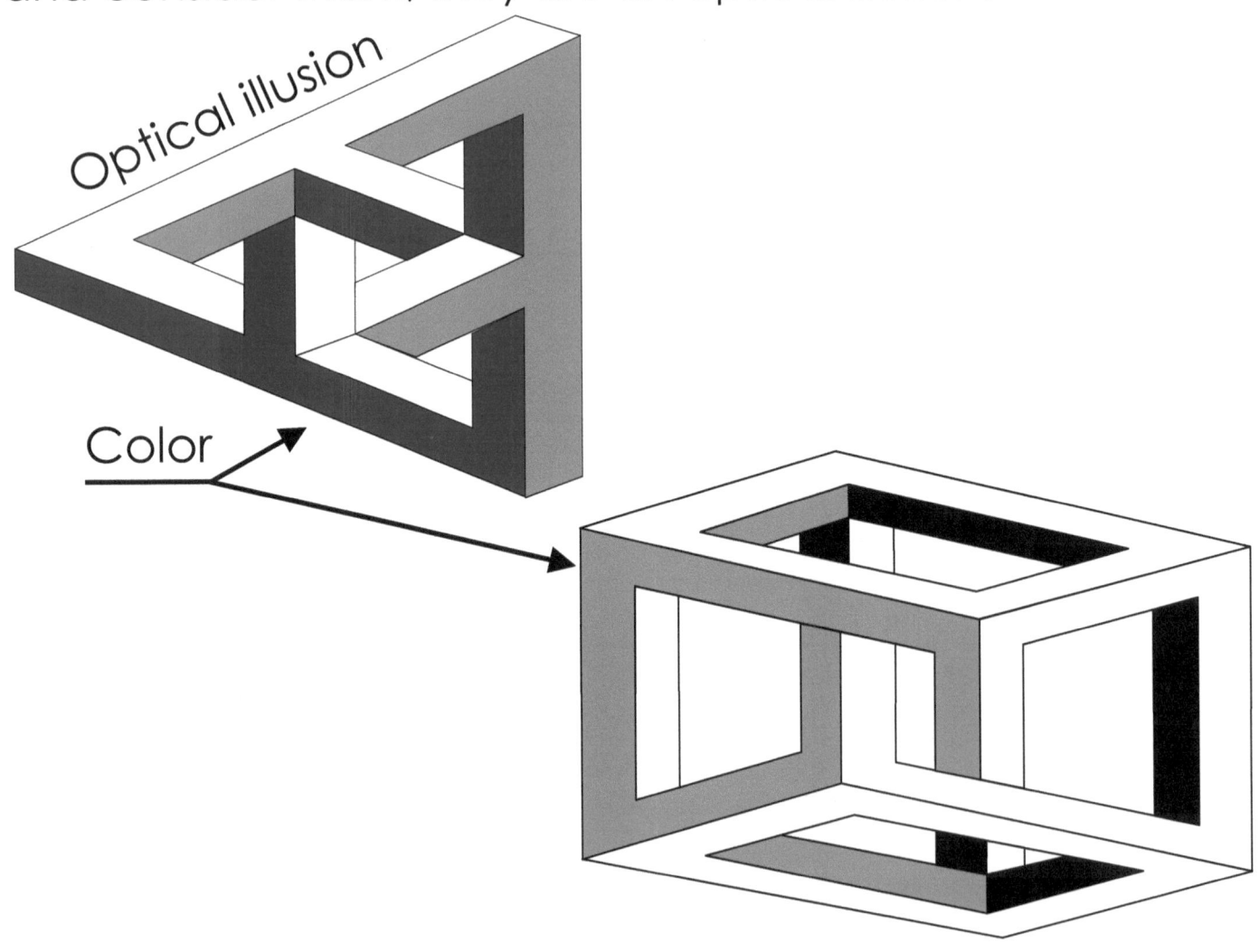

www.ingramcontent.com/pod-product-compliance
Lightning Source LLC
Chambersburg PA
CBHW050834180526
45159CB00004B/1897